3 Look at this photograph. You can find **1** adult bear and **2** bear cubs. How many bears are in the picture altogether? Look for groups of **3** throughout *Bears*.

Brown bear and cubs

Bears!
And very bear-y.

Bears! Bears, Bears,

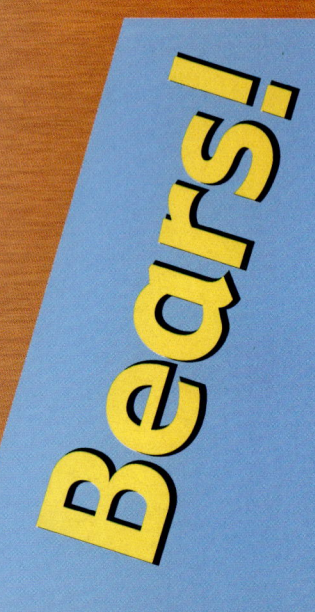

Giant pandas live in bamboo forests in China. Pandas have to munch bamboo to stay so roly-poly.

The **sun bear** is the world's smallest bear.

North American **black bears** are very curious. They're very good at climbing, and perch—or even sleep—in trees.

The **spectacled bear** looks like it's wearing spectacles, or eyeglasses!

Polar bears live in the Arctic, where it's icy and cold. Their blubber, thick fur, and black skin keep them warm.

Asiatic black bears are known as moon bears because of the moon-shaped markings on their chests.

Brown bears, including grizzlies, are big and strong. They hunt, fish, and forage for fruits, nuts, and insects.

Furry **sloth bears** suck termites right out of their nests! Yum!

Food for Three
Which One's For Me?

3

There are **3** bears in this puzzle. Can you find them? What kinds of bears are they?

There are **3** foods in this puzzle, too. Can you help each bear find its way through the maze to the food that it likes best?

Answer Key:
Polar Bear = Seal
Panda = Bamboo
Asiatic Bear = Termites

Illustrations by Linda Howard Bittner

Little Cubs
Learning and Growing

How do little cubs make their way in the world? With a little help!

Brrr! Polar bear cubs are usually born in the frigid months of December and January. How do they stay warm in the icy Arctic? They snuggle against their furry moms for a cozy snooze in the snow.

Snuggly Bears

Rough 'n' Tumble

As they get older, bear cubs of all kinds—like these Asiatic black bears—learn to hunt and play by wrestling with their brothers and sisters. Grrr!

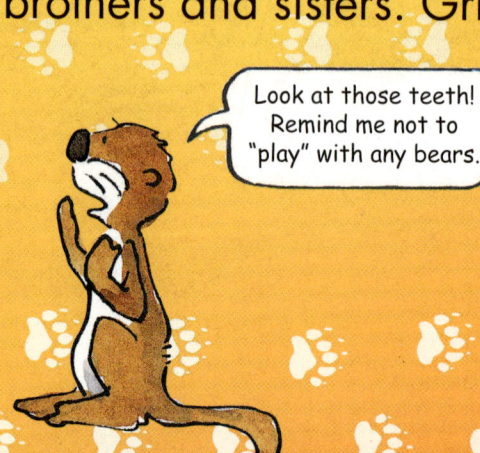

Look at those teeth! Remind me not to "play" with any bears.

6

Hairless and Helpless

Like all bear cubs, even "giant" pandas are born tiny—smaller than a grown-up's hand! They depend on their mothers to keep them warm, safe, and fed.

Same Different

Same means alike. **Different** means not alike. Some of these bears have the **same** colors. Can you find them? Which bears have **different** colors?

Gone Fishin'

Mother grizzly bears teach their cubs to hunt, fish, and forage, so when they get bigger, they can feed themselves.

The Bear Burrows In

By Sarah Hoban

Stacking sticks

With crackling leaves,

The brown bear makes his bed.

He gobbles nuts

And roots and fruit

To face the cold ahead.

Cozy,

Dozing in his den

And snuggled head to toe,

The brown bear naps

And barely hears

The windy winter blow.

Art by Toby Williams

9

How Now, Big Bear?

How big are bears?

Bears come in lots of different sizes. Some bears, like the **panda**, stand about 5 feet tall. That's a little taller than a fifth grader.

Now there's one who's MY size!

Others, like the **grizzly**, tower at nine feet. That's taller than a professional basketball player!

10

How **furry** are bears?

All bears have thick, coarse fur, but each bear's fur is a little different.

To stay warm in the Arctic, polar bears have two layers of fur. The top layer is waterproof!

Sloth bears have longer, shaggy black fur that sticks out from their bodies and around their necks.

How **ferocious** are bears?

Most bears stay away from people if they can, and some would rather fill up on plants, berries, and insects than go hunting.

But some bears—like **grizzlies**—are aggressive if confronted, protecting their cubs, or looking for food.

Same Different

In some ways, all bears are the **same**. They all have four legs, thick fur, big snouts, and big appetites! In some ways, they are **different**. They live in **different** places, eat **different** things, and are **different** sizes.

Look at these pictures. What are some things that are the **same** about *all* of these bears? What are some things that are the **same** about *some* of the bears? Now, what are some things that are **different**?

11

Barry's Very Grown-up Day

by Linda L. Covella

Mama Black Bear wakes Barry with a gentle nudge. Springtime! After seven months of hibernation, Barry is ready to leave the den. Today he feels very grown up. Only one year ago, he was a helpless, hairless baby. Brrrr! Now, thank goodness, he has his own fur to keep him warm. Mama says he must learn to feed himself because in a few months, he'll be on his own.
She coaxes him toward the opening. Time for his first lesson.

illustrations by George Angelini

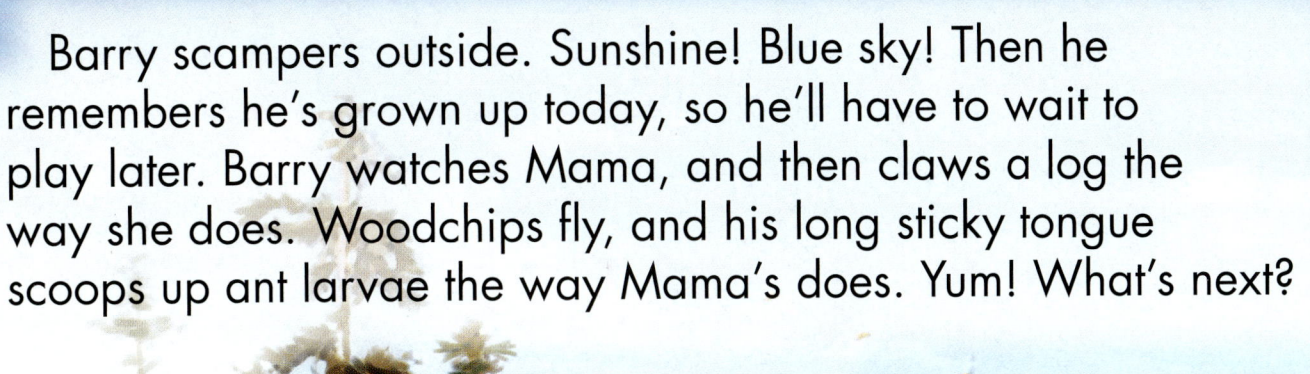

Barry scampers outside. Sunshine! Blue sky! Then he remembers he's grown up today, so he'll have to wait to play later. Barry watches Mama, and then claws a log the way she does. Woodchips fly, and his long sticky tongue scoops up ant larvae the way Mama's does. Yum! What's next?

Mama stands and sniffs the air. Barry smells it, too. Honey! Mama pulls a honeycomb out of a tree hole. Bees swarm. They cling to Mama and Barry. A bee sinks its stinger into Barry. Ouch! Copying Mama, he shakes his fur, and throws the bees off like a spray of water. He runs after Mama to share the prize: bee larvae and honey!

Mama must find more food. Phew! Tired! Barry's learned enough for one day. Mama hides him under a bush, tells him to stay low and quiet, and then she lumbers off into the woods. Barry is thirsty after all his hard work. The sparkling stream is so close. It would only take a minute . . .

Mm, the water is cool and fresh. Suddenly, Barry's round ears perk up and he peers through the woods. Humans! He clacks his teeth in fear. Oh, where's Mama?

Mama returns. She grunts. Hurry! Barry runs and climbs a tree. Mama follows. From the tree, they watch the humans pass. Barry snuggles close to Mama and looks down at his world. When the time comes, he'll be ready to live on his own. For now, he's happy to be with Mama.

Keeping Bears Busy

By Rachel Young

In the wild, bears spend all day looking for food and staying safe. But in zoos, people bring bears food and make sure they're healthy. That's why zookeepers invent all sorts of games and puzzles to keep zoo bears busy, active, and having fun—and keep their hunting, climbing, and smelling skills sharp, too!

Keepers hide "snacks" in bears' habitats. The bears have to sniff, climb, and "hunt" for their treats the way they would in the wild!

You Can Be a Scientist, Too!

Bears aren't the only animals that like to learn new games. Try making these simple toys for your pets.

Dogs' ancestors had to find their own food. Hide healthy dog treats around the house or inside a toy for a little doggy hide-and-seek.

Tie one end of a piece of string to a feather or scrap of fabric, and tie the other end to a stick. "Cast" the toy as if you're fishing, then reel it back. Your cat can practice her wild-cat pounce!